THE MAGNIFICENT GIANTS OF THE OCEAN

OMAK CHARLIE OMAR

*To Agnes, Anthony, George,
Clement, Helen,
Jacob, Michael, Stella,
Damian, Mercy, Sharon, Blessing & YOU.*

Copyright ©2024 O. Charles O.

All rights reserved. No part of this book may be reproduced or transmitted in any form or by any means without written permission from the author, except for the inclusion of brief quotations in a review

Contents

Preface...........vii

Introduction....................................1

The Blue Whale................................7

The Sperm Whale.............................17

The Fin Whale.................................25

The Orca..33

The Humpback Whale........................43

The Whale Shark..............................51

The Giant Squid...............................61

The Great White Shark.......................69

The Leatherback Sea Turtle..................77

Contents

The Manta Ray..............................87

The Greenland Shark......................95

The Future of the Ocean Giants..........103

Glossary..109

References....................................115

Preface

Humans have long been captivated by the oceans, vast and mysterious, hiding worlds teeming with life. Beneath the waves, some of our planet's most extraordinary creatures dwell—the ocean giants: powerful, and vital for the stability of aquatic ecosystems. From the blue whale to the giant squid, these creatures have evolved to thrive in environments that even our most advanced machines struggle to reach.

Throughout history, these ocean giants have inspired legends, scientific discoveries, and stories. They represent the peak of evolution within their unique habitats, showcasing both the beauty and ferocity of deep-sea life. Yet despite their size and strength, many of these giants are surprisingly vulnerable. Climate change, overfishing, and habitat disruption pose serious threats to their survival, making it more critical than ever to understand and protect them.

This book invites you to explore the lives of these ocean giants. Each chapter delves into the anatomy, behaviors, and unique characteristics of these remarkable creatures, unveiling the complex roles they play in their ecosystems. You'll learn how the blue whale's heart powers its lengthy migrations, how the sperm whale navigates the ocean's depths, and how the leatherback sea turtle undertakes one of the ocean's most astonishing migrations.

Preface

Yet this is more than a book on biology; it's also an exploration of our relationship with these giants. Historically, we've hunted them, but now we are called to be their protectors. As you read, consider the challenges these animals face and the ongoing efforts to ensure their survival for generations to come.

Whether you're young and curious, a nature enthusiast, or simply someone interested in the natural world, this book offers a deeper appreciation for the ocean's giants. These creatures are more than just sea dwellers—they embody the planet's resilience and diversity. Understanding them is a crucial step in preserving the delicate balance of life on Earth.

1

Introduction

The vastness and mystery of the ocean have enabled its giants to not only survive but also become major elements of the marine food chain. Due to their massive size, these creatures always find themselves at the top, playing an important role in keeping their ecosystems balanced. One of these giants, the blue whale, is one of the most iconic. Blue whales can grow to incredible lengths and weigh several tons, dwarfing even the largest dinosaurs. Yet, what truly defines them is not just their size but also the role they play in the ocean's health.

A blue whale's daily consumption of tons of krill regulate krill populations, balancing the marine ecosystem. Their migration and feeding patterns also play an important role in nutrient cycling through a process called "whale pump," where their feces fertilizes the ocean, promoting phytoplankton growth—the foundation of the marine food chain. But blue whales aren't the only sea giants with a massive impact.

The whale shark is another fascinating example. Although not a mammal like the blue whale, the whale shark shares similar feeding habits. Growing up to 40 feet long and weighing around 20 tons, it filters plankton and small fish from the water. Despite this fearsome size, the whale shark is gentle and poses no threat to humans, making it a favorite among divers. Its peaceful nature and unique feeding adaptations show what it means to be an ocean giant— dominating not with force but through a vital ecological role.

Another characteristic that sets ocean giants apart is their longevity. Many of these creatures live longer than other marine species, making them keystone species in their ecosystems. The Greenland shark, for example, is one of the largest sharks, measuring up to 21 feet in length, but it also holds the title of the longest-living vertebrate, with some estimates placing its lifespan at nearly 500 years. This extraordinary longevity makes it a living archive of ocean history, having witnessed centuries of change in the marine environment.

Similarly, the bowhead whale, another Arctic giant, can live for over 200 years. Its thick blubber allows it to survive in some of the world's harshest environments, and its long life contributes to the stability of its ecosystem, as it maintains its position as a critical species in Arctic food chains. These creatures' long lifespans raise questions about aging

Introduction

and survival in extreme conditions, making them indicators of ocean health.

The evolutionary journey of these giants is equally fascinating. Ocean giants did not always exist as the massive creatures we know today. Many, like whales, started their evolutionary path on land. Around 50 million years ago, terrestrial mammals known as Pakicetids began gradually adapting to aquatic life. Over time, their limbs evolved into flippers, their bodies became streamlined for swimming, and their nostrils moved to the tops of their heads, becoming blowholes that allowed them to breathe while swimming.

Sharks, often seen as the sea's apex predators, also have an evolutionary history that predates dinosaurs. One of the most famous prehistoric giants is the megalodon, which could grow up to 60 feet long and dominate the oceans, preying on large marine mammals. Although the megalodon went extinct millions of years ago, its legacy endures in modern sharks that continue to rule the seas.

The adaptation of other species, such as marine reptiles like Ichthyosaurs, further demonstrates the diversity of ocean giants. These sleek, dolphin-like predators evolved powerful tails for swimming and grew to sizes comparable to today's whales. Meanwhile, the ancestors of modern sea turtles transitioned from land to sea, growing flippers for long-distance ocean travel. The leatherback sea turtle is a testament to this evolutionary success, as it continues to journey thousands of miles across the ocean each year.

As ocean giants evolved, they developed physical adaptations and advanced social behaviors and communication. Cetaceans like dolphins and orcas are known for their sophisticated communication systems, which include clicks, whistles, and body language. These communication methods are critical for hunting, navigating, and maintaining social bonds within their pods. This ability to coordinate within groups has given these intelligent creatures a significant evolutionary advantage, allowing them to thrive in complex marine environments.

The presence of these ocean giants often inspires admiration and reverence. Throughout human history, they have held significant cultural importance, symbolizing the mysteries and power of the sea. From the Maori legend of the whale as a guardian spirit to the Norse myth of Jörmungandr, the Midgard serpent, these creatures have captured the human imagination. In modern times, this reverence has evolved into a movement to protect these giants and their habitats. Their survival is essential to maintaining ecological balance and reflects humanity's responsibility to preserve the natural world for future generations.

Thus, while size is certainly a defining feature of ocean giants, their impact on the environment, their

Introduction

long lives, their evolutionary adaptations, and their symbolic importance to humans truly set them apart. These giants are more than just creatures of immense size; they are integral to the ocean's ecosystems and hold a special place in the natural world and human culture.

2

The Blue Whale

The blue whale, the largest animal on Earth, symbolizes nature's grandeur and resilience. Reaching lengths of up to 100 feet and weighing as much as 200 tons, these majestic creatures dwarf the largest dinosaurs. Even their body parts are huge; a blue whale's tongue can weigh as much as an elephant's, and its heart is the size of a small car. Yet, despite their massive size, blue whales move with an elegance that belies their enormous bulk, gliding effortlessly through the world's oceans.

This sheer size is the result of millions of years of evolution. Blue whales have adapted into the ultimate oceanic giants, perfectly designed to travel long distances for food and mates. Their size also allows them to endure the harsh environments of the open seas, a world where food is scarce, and the challenges of survival are extreme. Among the most interesting aspects of their nature is that they rely on some of the ocean's smallest organisms for survival—krill.

Despite their giant size, blue whales feed almost exclusively on krill, tiny shrimp-like creatures found in large swarms in ocean regions. During feeding season, a single blue whale can consume up to 4 tons of krill per day, highlighting the energy demands of their massive bodies. Their feeding strategy, known as lunge feeding, is a marvel of evolutionary efficiency. Blue whales will accelerate toward a swarm of krill with their mouths open, engulfing vast quantities of water and prey. Afterward, they push the water out through their baleen plates—comb-like structures made of keratin—leaving the krill trapped inside, ready for consumption.

This dynamic between the largest and some of the smallest ocean creatures reminds us of the delicate balance within marine ecosystems. The success of the blue whale is tightly linked to the abundance of krill, showcasing a unique ecological connection that also ties their survival to the ocean's health.

Historically, the size of the blue whale has been a subject of fascination and exploitation. Whalers of the 19th and early 20th centuries often described the overwhelming sight of these colossal creatures surfacing from the depths, their massive bodies seeming otherworldly against the vast ocean backdrop. The first sign of a blue whale's presence would often be its blow—a misty column of water shooting up to 30 feet into the air. Yet, these creatures were hunted almost to extinction for their valuable blubber, used to produce oil. By the time commercial whaling was banned in the 20th century, blue whale populations had been decimated by as much as 99%.

Blue whales remain with us today thanks to international conservation efforts and the species' resilience. However, their populations are still only a fraction of what they once were. The conservation success stories, however, provide hope. Through protective legislation, such as bans on commercial whaling and establishing marine protected areas, blue whales have slowly recovered in some parts of the world. These efforts underscore the importance of preserving our oceans and their incredible creatures.

In addition to their size and survival, blue whales possess another remarkable trait: their vocal abilities. They produce some of the loudest sounds of any animal on Earth, with their calls reaching up to 188 decibels—louder than a jet engine. These low-frequency sounds, which travel hundreds of miles underwater, serve as a vital tool for communication, especially in the vast and often desolate ocean. The songs of blue whales are essential during the mating season and for navigating through the deep waters with limited visibility.

Blue whale communication is critical not only for social interaction but also for survival. Their long, low moans and pulses can travel across ocean basins, helping them maintain contact with other whales despite their solitary nature. These creatures often roam the ocean alone, and only gather to breed or feed. Still, their calls allow them to stay connected across great distances. This ability to communicate over vast areas is an adaptation perfectly suited for life in the open ocean's expansive and often isolating environment.

Their migratory behavior also demonstrates the blue whale's extraordinary endurance and strength. Each year, these whales undertake migrations spanning thousands of miles, traveling between their feeding grounds in the Arctic's cold waters and breeding grounds in the warmer tropical and subtropical seas. This migration is a test of endurance and a survival strategy. By feeding extensively in Polar Regions during the summer months, blue whales build up fat reserves that sustain them during the long journey to their breeding grounds, where food is scarcer.

One famous example of blue whale migration is the story of "Big Blue," a blue whale tracked by researchers on an epic journey from California to Costa Rica. Big Blue traveled more than 5,000 miles for several months, navigating waters filled with potential predators and other natural obstacles. Tracking this whale gave us knowledge of the migration patterns and behaviors of blue whales, highlighting the importance of protecting the critical habitats they depend on for survival.

Despite their incredible abilities, blue whales remain vulnerable to many modern threats, most of which are human-caused. Although commercial whaling has been largely banned, blue whales now face new challenges like climate change, ship strikes, and entanglement in fishing gear. The warming of ocean waters, driven by climate change, threatens the abundance of krill, the blue whale's primary food source. Meanwhile, increased shipping traffic raises the risk of fatal collisions with these massive creatures. At the same time, fishing gear poses a threat of entanglement, which can lead to injury or death.

Noise pollution is another growing concern. Once a quiet realm where blue whales' calls could be heard across vast distances, the ocean is now filled with the sounds of human activity—ship engines, industrial operations, and military sonar. These sounds interfere with the whales' ability to communicate, navigate, and locate food. Research showed that blue whales reduce the frequency of their calls in noisy areas or even stop vocalizing at all. In regions like the North Pacific, where ship traffic is particularly high, blue whales have been observed changing their calling patterns, which could disrupt their natural behaviors.

One striking example of the impact of noise pollution on blue whale communication comes from the Santa Barbara Channel, off the coast of California. This busy shipping lane intersects with blue whale migration routes, and researchers have noticed a decrease in whale vocalizations during heavy ship traffic. This disruption in communication could have serious consequences for the species' ability to find mates, navigate, and avoid predators. The situation has prompted efforts to reduce the impact of noise pollution, including the development of quieter ship engines and the establishment of "quiet zones" where human activity is limited.

The story of "52 Blue," a blue whale that calls at an unusually high frequency of 52 hertz, further illustrates the challenges posed by noise in the ocean. This whale, often called the "world's loneliest whale," was first found in the late 1980s. Its call has never been matched by any other whale, leading scientists to speculate that it may be isolated from its peers, unable to communicate effectively with its own kind. While the exact cause of this whale's unusual call remains unknown, it highlights the importance of protecting the ocean to ensure the survival of species like the blue whale.

Despite these challenges, blue whales have shown remarkable adaptability. Some studies suggest that they are beginning to adjust the frequency of their calls in response to the increasing levels of noise in the ocean. By lowering the pitch of their vocalizations, blue whales may be attempting to communicate more effectively over long distances as lower-frequency sounds travel further underwater.

This behavior demonstrates their resilience and ability to evolve in response to environmental changes. However, it also highlights the urgent need for continued efforts to reduce human impacts on the ocean, including noise pollution.

Conservation initiatives have already made a difference. Establishing marine protected areas and quiet zones, where shipping and industrial activity are restricted, has positively affected blue whale populations. For example, in regions where noise levels have been reduced, scientists have observed an increase in whale vocalizations, suggesting that these efforts are helping to restore the natural acoustic environment of the ocean. Such measures are essential not only for the survival of blue whales but for the health of the entire marine ecosystem, as the calls of blue whales—sometimes referred to as the "heartbeat of the ocean"—play a role in regulating the behavior of other marine species.

The story of the blue whale is one of survival and adaptation in the face of immense challenges. From their evolutionary origins as the largest animals on Earth to their delicate relationship with the ocean's tiniest creatures, blue whales embody the intricate connections that sustain life in the ocean. Their vocalizations, migrations, and feeding strategies are a testament to the incredible adaptations they have developed over millions of years. However, the continued survival of blue whales depends on our ability to protect their habitats from the threats of climate change, noise pollution, and human activity. With concerted conservation efforts, there is hope that blue whales will continue to thrive in the world's oceans, inspiring awe and respect for future generations.

3

The Sperm Whale

The sperm whale is another marvel of deep-sea adaptation, with extraordinary hunting abilities that allow it to prosper in one of the world's harshest environments. Known for diving to incredible depths, sometimes exceeding 3,000 meters (nearly 10,000 feet), sperm whales pursue their preferred prey—the giant squid. These dives, which can last more than an hour, are a testament to the whale's remarkable physiology. By slowing their heart rate and storing oxygen, sperm whales excel as hunters. They are among the world's deepest-diving mammals.

Central to the sperm whale's hunting success is its use of echolocation to navigate the ocean's pitch-black depths and find prey. This sophisticated biological sonar system allows the whale to emit powerful clicks that bounce off objects, returning as echoes that create a detailed "image" of the surroundings. The clicks of a sperm whale are also among the loudest sounds made by any animal. They are so precise that the whale can detect a squid as small as a meter long from hundreds of meters away. This ability allows sperm whales to thrive in environments without light, relying entirely on sound to "see" their prey.

Evidence of the sperm whale's hunting prowess can be seen in the battle scars found on many of these whales caused by giant squid's tentacles. These scars speak to the epic underwater battles between predator and prey. Tales from nineteenth-century whalers frequently described clashes between sperm whales and giant squid, with even the largest squid struggling to escape the pursuit of a whale. These stories, alongside modern knowledge, paint a vivid picture of a creature perfectly suited for deep-sea hunting, equipped with both physical and sensory tools to dominate one of the most difficult environments on the planet.

Yet, the sperm whale's hunting strategies go beyond physical prowess and include complex social behaviors that reveal a high level of intelligence. Sperm whales often hunt in groups called pods, where cooperation is essential to their success. These pods, typically consisting of females and their young, work together to locate and corner prey, particularly in areas where food is abundant. The coordinated efforts of a pod resemble a pack of wolves on land, with each whale playing an important role in the hunt. This social structure improves hunting efficiency and ensures the education of younger whales who observe and participate in group hunts.

A fascinating element of their social hunting behavior is their use of "codas"—click patterns believed to be a form of communication, potentially conveying information about prey or coordinating movements during a hunt. Researchers have discovered that different pods use distinct codas, almost like dialects, suggesting that these vocalizations hold functional and cultural significance. A pod off the coast of New Zealand may use a different set of codas than a pod in the North Atlantic, showcasing the diversity and complexity of sperm whale communication. This aspect of hunting behavior broadens our understanding of these magnificent giants and their intricate social structures.

The cultural transmission of hunting techniques within pods demonstrates how knowledge is passed down through generations to ensure survival in different regions. In the 1960s, a group of sperm whales in the Pacific Ocean was observed using a new hunting technique: they would dive beneath a school of squid and blow bubbles to disorient their prey, making it easier to catch. This behavior, unique to this population, exemplifies the adaptability and creativity of sperm whales. It also implies that these whales can develop and refine hunting strategies based on their specific environment, passing these techniques down to their offspring and ensuring their way of life continues in the deep ocean.

Remarkable as their hunting techniques are, the sperm whale's survival in the deep ocean is made possible by unique physiological adaptations. Their ability to dive for up to 90 minutes is aided by a flexible ribcage that allows their lungs to collapse under pressure, reducing the risk of decompression sickness. Additionally, their large, oil-filled heads help regulate buoyancy and focus echolocation clicks—both critical for finding prey in the ocean's lightless depths.

One particularly striking example of the sperm whale's deep-sea prowess occurred in the 1980s when researchers encountered a male sperm whale with deep scars believed to be from a battle with a giant squid. These scars showed evidence of the fierce confrontations in the abyssal depths, where these two titans of the deep engage in life-or-death battles. The marks left by the squid's suction cups and hooks on the whale's skin are a testament to the intensity of these battles. These tales highlight the sperm whale's role as a top predator and add to the mystique surrounding these giants of the deep, whose lives remain largely hidden from human eyes.

Another aspect of their social behavior is "alloparenting," where pod members take turns caring for the calves. At the same time, the mothers dive deep for food. This cooperative care is crucial for the young's survival, as feeding requires dives that take mothers away for long periods.

Despite their reputation as solitary hunters, sperm whales are highly social creatures that form tightly knit pods. While adult males often live solitary lives, joining pods only during the mating season, females and their young form complex matriarchal groups. These pods exhibit strong bonds through regular physical contact, vocal communication, and coordinated activities like caring for newborns.

When faced with threats, sperm whales cooperate to protect vulnerable members, such as calves or injured individuals. In such instances, the pod will form a defensive circle, known as a "marguerite formation," with heads facing inward and tails outward.

This formation allows them to fend off attackers, like

orcas, with powerful tail slaps. A notable story tells of a pod in the North Atlantic successfully defending its young from a pod of orcas using this technique. This social organization protects the young from predators like orcas and teaches them essential survival skills, like diving and hunting.

One intriguing example of cooperative behavior happened off the coast of New Zealand, where researchers observed sperm whales diving in synchrony. Although male sperm whales are known to hunt individually, this group engaged in cooperative hunting. By diving together, they could herd schools of fish into tighter clusters, making it easier for individual whales to catch prey.

Beyond hunting, sperm whales also participate in social interactions that reinforce group cohesion. Communal resting, known as "logging," is one such practice. During these sessions, sperm whales float motionless on the ocean's surface, usually side by side, resembling logs of wood drifting in the water. Logging not only allows the whales to rest but also promotes social bonding among pods. Observations of logging pods reveal that whales often engage in gentle fin touches, reinforcing their social connections. Even in their most relaxed states, sperm whales remain connected, nurturing the relationships essential to survival. This behavior exemplifies their intelligence, unity, and the strong social bonds that define the enigmatic sperm whale.

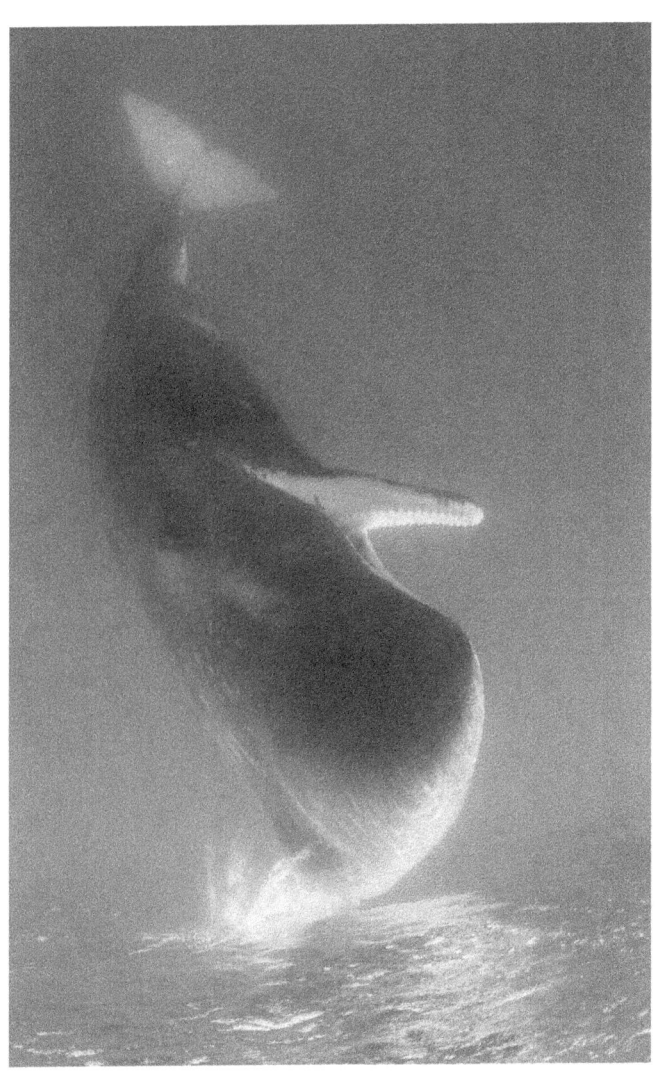

4

The Fin Whale

The fin whale, sometimes called the "greyhound of the sea," is renowned for its incredible speed and agility. These sleek giants can reach up to 23 mph, making them the fastest whale species in the ocean. Their streamlined bodies are perfect for rapid movement, allowing them to travel long distances during their migratory journeys. One fascinating example is the fin whale's migration from cold, nutrient-rich Arctic waters to warmer breeding grounds closer to the equator. These journeys can be thousands of miles long. Still, the fin whale's incredible speed allows it to easily complete them, ensuring it arrives on time for the breeding season.

The fin whale's unique body structure contributes significantly to this impressive speed. Unlike the bulkier blue whale, the fin whale's body is slender and shaped like a torpedo, reducing drag as it slices through the water. This design is enhanced by long, narrow pectoral fins and a powerful tail fluke, which propel the whale forward with every stroke. Their movement efficiency allows them to maintain high speeds for long periods, which is essential during migration and for avoiding potential predators. In one case, a fin whale was seen outpacing a group of orcas, showing how fast it can move, and its resilience in the face of potential danger.

This speed and agility are more than just physical accomplishments; they also play an important role in the fin whale's feeding strategy. Unlike other large whales, which rely on slow, methodical filter feeding, the fin whale uses a technique called "lunge feeding." This involves the whale accelerating toward a krill or small fish school before opening its mouth to swallow much water and prey. Additionally, their asymmetrical jaws, unique among whales, with the right side lighter in color than the left. Scientists believe this asymmetry may aid fin whales' hunting strategy by disorienting prey or improving their vision while feeding. This speed allows it to surprise its prey, and its large, pleated throat expands to swallow as much food as possible. After catching its prey, the whale closes its mouth and uses its baleen plates to filter out the water, leaving behind a large concentration of food.

The adaptability of the fin whale in its feeding strategy is remarkable. Unlike species specializing in a specific type of prey, the fin whale is an opportunistic feeder, meaning it can adjust its diet based on food availability. During periods of high krill abundance in the Polar Regions, fin whales feast on these tiny crustaceans, taking advantage of the seasonal bounty. When krill become scarce, they shift their diet to include small fish, squid, and other marine organisms, demonstrating their survival ability in various environments. This adaptability is critical in the vast and often unpredictable ocean, where prey populations can shift dramatically due to environmental changes.

Another remarkable fin whale adaptability is their ability to feed at various depths. While fin whales are frequently seen feeding near the surface, they can also dive to great depths in search of food. These deep dives, which can reach over 200 meters (656 feet), allow them to access prey that many other predators cannot. Researchers observed a fin whale repeatedly diving to the ocean floor to feed on dense schools of fish that surface-feeding marine animals couldn't reach. This behavior illustrates the whale's versatility and understanding of the ocean's complex ecosystems, where survival often depends on the ability to apply multiple feeding strategies.

Fin whales' social behavior also contributes to their survival in the oceans. While they are usually seen in small groups, often in pairs or trios, larger groups can form during feeding or migration. These social structures are flexible, with individuals coming together or separating depending on food availability or the time of year. This flexibility helps fin whales to increase their chances of survival, especially in environments with uneven resource distribution. During the summer in the North Atlantic, large groups of fin whales gather on feeding grounds rich in krill and small fish. These gatherings help in efficient feeding and strengthen social bonds as whales also engage in cooperative behaviors like synchronized diving and bubble net feeding.

Fin whales use low-frequency vocalizations to communicate across long distances. These vocalizations, often described by researchers as long, deep moans, can also travel hundreds of miles underwater, allowing fin whales to communicate across the ocean. Like other whales, this communication is essential for organizing group activities. During the annual migration from Arctic feeding grounds to warmer breeding areas, these vocalizations keep fin whales connected to one another, allowing them to remain part of a larger social network. In one case, a group of fin whales was observed using these calls to coordinate their movements across a large region of the North Atlantic, demonstrating the importance of vocal communication in their social structure.

Another example of the fin whale's resilience and adaptability comes from its interactions with human activities in the ocean. Historically, whalers heavily hunted fin whales because of their large size and high yield of oil and baleen. Despite intense hunting pressure significantly reducing their populations, fin whales have shown remarkable resilience in some areas.

Fin whales' resilience is best demonstrated by their slow but steady recovery from the brink of extinction. After centuries of persistent hunting, fin whale populations were severely depleted, with some estimates indicating that only a small fraction of their original numbers remained. Fin whale sightings have increased in recent years in areas where they were previously rare, proving their ability to recover from challenging circumstances. This recovery is a success story for the species and a stark reminder of nature's ability to heal when given the opportunity.

The story of the fin whale's resurgence is one of hope and a call to protect these magnificent creatures so that future generations can see their grace and majesty in the ocean. Their adaptability, physical prowess, and resilience testify to their evolutionary success in an ever-changing ocean environment.

5

The Orca

Orcas, also known as "killer whales", are the ocean's undisputed apex predators, ruling the seas with intelligence, strength, and one of the most sophisticated social cooperations. Unlike most other marine predators, orcas do not have natural enemies, allowing them to hunt freely and dominate a variety of marine environments. Their status as apex predators is due to their physical prowess and complex hunting strategies, which rank among the most complex in the animal kingdom.

One of the most incredible parts of orca behavior is their ability to hunt together in pods. These highly social animals live in tightly connected family groups that can last multiple generations. Orcas in these pods use coordinated hunting techniques passed down from generation to generation. Like the one in the waters of Patagonia, a particular group of orcas is known for a hunting method called "beach hunting." These orcas deliberately strand themselves on the beach to catch seals, a risky action that needs precise timing and coordination. This strategy shows the orca's incredible versatility and ability to learn and teach complex behaviors.

Another example of orca hunting intelligence is found in the waters of Norway, where orcas have been seen using a technique known as "carousel feeding." Orcas herd herring into tight balls by swimming in circles and flashing their white undersides, confusing the fish. When the herring is packed tightly, the orcas use powerful tail slaps to stun them before eating them. This behavior demonstrates the orca's ability to manipulate its environment and prey, using physical and psychological tactics to secure food. These instances demonstrate the orca's supremacy as an apex predator and their remarkable ability to adapt their hunting strategies to varying environments and prey.

Along with their advanced hunting techniques, orcas have an excellent knowledge of their prey species' social structures, increasing their success as predators. Particularly in Antarctica's icy waters, orcas have been seen preying on Minke whales by targeting the weakest pod member, usually a calf or weak elderly whale. The orcas work together to separate the vulnerable whale from its group, reducing the risk of injury and increasing the chances of a successful hunt, just like pride of lions hunt in groups. This strategic behavior demonstrates the orca's ability to assess and exploit the social structure of their prey, which is rare among marine predators.

Orcas are also known to engage in social behaviors, with different pods having unique hunting techniques and customs. The bonds within orca pods are so strong that they span generations, with multiple generations living in tightly connected groups. These multi-generational pods typically comprise a matriarch and her descendants, resulting in a decades-long stable social unit. The matriarch is an important guide for the pod. She uses her experience to lead the group in finding food, navigating migration routes, and avoiding dangers. Southern Resident killer whales are one example of this, with matriarchs leading their pods through familiar routes and ensuring the safety of younger members. This matriarchal leadership emphasizes the younger orcas' deep respect and trust for their elders, reinforcing the pod's family bonds.

One example is Granny, an orca matriarch thought to have lived for over a century. Granny was well-known for leading her pod through the waters of the Pacific Northwest, teaching younger members important survival skills. This longevity and leadership role highlights the significance of elder orcas in passing on knowledge and upholding the pod's cultural traditions. The tightly connected structure of orca pods, in which family members live together for life, demonstrates their advanced social intelligence. These social behaviors within orca pods are fascinating, with relationships often characterized by deep emotional connections and cooperative behaviors.

One of the most fascinating aspects of orca intelligence and adaptability is their ability to communicate complex information via vocalizations. Like whales, orcas make various calls, clicks, and whistles to coordinate hunts, maintain social bonds, and navigate the ocean. Each orca pod has a unique dialect passed down from generation to generation. This vocal learning demonstrates the orca's brain power, as it requires memory, learning, and adapting to new circumstances. In some cases, orcas have been seen mimicking the sounds of other marine animals or human-made noises, demonstrating their vocal range and intelligence. These vocalizations are more than just noises; they are a sophisticated communication essential to the orca's survival as an apex predator.

These variations in social structure are especially noticeable when comparing orca populations worldwide. Resident orcas primarily feed on fish, particularly salmon, in the North Pacific and have developed specialized hunting techniques for catching prey. Transient orcas in the same region, on the other hand, prefer marine mammals like seals and sea lions and have developed stealthier and more ruthless hunting methods. Despite their proximity, these behavioral variations suggest that orcas, like humans, can learn and transmit knowledge within their pods.

Orca pods' social structures extend beyond just familial bonds to include a complex division of roles that ensures the group's survival. Individual orcas in many pods play specific roles, such as primary hunters, protectors, or caregivers. This division of labor is especially evident during hunting when the pod uses precise coordination and cooperation strategies. In addition to cooperative hunting and role specialization, orcas demonstrate empathy and caregiving behavior within their pods. Orcas' care for sick or injured pod members is frequently reported, demonstrating their strong sense of responsibility.

In one case, an adult orca was seen assisting a sick pod member to the surface for air and remaining by its side until it recovered. This behavior demonstrates not only orcas' strong social bonds but also their emotional intelligence. How they care for one another, especially in times of need, shows the strength of their social connections and the value of family and community.

Tahlequah, a Southern Resident orca who made headlines in 2018, is one example of how strong orca social bonds can be. Tahlequah gave birth to a calf who tragically died about 30 minutes after birth. In her expression of grief, she carried the calf's body on her head for 17 days, traveling more than 1,000 miles. This heartbreaking event drew global attention and highlighted the emotional complexity of orcas. Stories like this show orcas' complex emotional lives and deep familial connections, pointing out their status as one of the most socially complex animals in the ocean.

Orcas also have a profound influence on the ecosystem. Orcas help to keep marine ecosystems balanced by controlling the populations of their prey. Conversely, the decline of sea otter populations in some areas has been linked to orca predation, which caused uncontrolled growth of sea urchin populations, damaging kelp forests. This chain of events demonstrates how the presence or absence of orcas can drastically change the marine ecosystem. As a result, orcas are more than just dominant hunters; they also play an important role in maintaining the state of the marine ecosystem. Their impact on the marine realm is crucial, placing them among the most important species in the ocean's intricate web of life.

One other fascinating aspect of orca behavior is their playful nature, observed both in the wild and in captivity. Orcas are known to surf waves, toss prey into the air, and even approach boats. These actions are not purely for amusement; they help younger orcas practice hunting skills and strengthen social bonds within the pod. Off the coast of Patagonia, for instance, scientists have observed orcas playing with seaweed and mimicking prey movements—behaviors that demonstrate their creativity and problem-solving abilities, showing the intelligence of these apex predators.

Despite their playful side, orcas have a complicated relationship with humans, especially in captivity. Orcas have been trained for entertainment and sports, performing acrobatic stunts in marine parks worldwide. However, captivity can sometimes lead to aggression. In 2010, tragedy struck at SeaWorld Orlando when an orca named Tilikum fatally attacked his trainer, Dawn Brancheau, during a live performance. This tragic incident highlighted the risks associated with keeping orcas in confined environments, sparking a conversation on the ethical and safety concerns surrounding their captivity.

In the wild, orcas rarely display aggression toward humans and have never been known to attack humans fatally. A notable incident occurred in 1972 off the coast of Point Sur, California, when an 18-year-old surfer, Hans Kretschmer, was bitten by an orca, likely mistaken for a seal. The orca quickly released the surfer, who survived the encounter but suffered significant injuries to the leg that required about 100 stitches to close the wounds. Such rare incidents reinforce that while orcas are curious about humans, they typically pose no threat in their natural habitats.

6

The Humpback Whale

The humpback whale, one of the most popular and loved whales, admired for its massive size and powerful presence. Adult humpback whales can grow to 60 feet long and weigh up to 40 tons, making them one of the world's largest creatures. Despite their massive size, they are known for their quickness in the water, sometimes surprising people watching with their ability to breach. The humpback whale's incredible breaching behavior is one of the most exciting demonstrations of its strength and size.

Humpback whales are known for their unusual and often mysterious behaviors - breaching, where the whale pushes its body out of the water and crashes back down, causing a huge splash. This amazing display, seen from miles away, has captivated people for ages. While the exact purpose of breaching is still being debated, it serves various purposes, including communication, play, and even parasite removal. Breaching may also allow whales to communicate with others, particularly during the breeding season. The sheer energy and coordination required to lift such a large body out of the water demonstrates the gentle giants' physical strength and agility.

Humpback whales make one of the longest annual migrations of any mammal, covering up to 5,000 miles from their feeding grounds in the Arctic to their breeding grounds in tropical waters, and from Antarctica's cold waters to the warm, shallow waters of Australia's Great Barrier Reef. During these migrations, they use their powerful tails, known as flukes, to push themselves across the ocean. These flukes can generate enough force to propel their massive bodies at speeds of up to 16 miles per hour - 7 mph short of the fin whale's speed, but still fast. The endurance and strength required for such journeys show the humpback's physical capabilities. During migration, whales face several challenges, including orca predation, shipping traffic, and changing sea temperatures. Yet, their enormous size and stamina allow them to overcome these challenges and complete one of nature's most breathtaking journeys.

The humpback whale's sheer physicality is matched by the strength of the bonds it forms, particularly between mothers and calves. A mother humpback whale fiercely protects her calf, frequently guiding and supporting her baby through the ocean's treacherous waters. One well-known story is about a humpback mother in the Pacific Ocean who was seen fighting off a pod of orcas that had targeted her calf. Using her enormous size and strength, the mother successfully fended off the orcas, ensuring her calf's safety. This display of maternal strength and commitment exemplifies the humpback's protective instincts and the strong bonds they form with their social groups.

Humpback whales also engage in another fascinating behavior: they use "bubble nets" to catch prey. This unique hunting technique involves a group of humpback whales forming a circular "net" of bubbles that traps schools of fish. The whales then swim up through the center of the net, mouths wide open, swallowing large amounts of prey in a single gulp. This cooperative feeding strategy showcases humpbacks' intelligence and social nature since it requires precise coordination and timing among pod members. This behavior occurred in the Gulf of Alaska, where a pod of humpback whales was seen repeatedly using bubble net feeding to successfully capture herring.

Humpback whales exhibit another intriguing behavior known as tail-slapping or lobtailing. The whale lifts its massive tail fluke out of the water and slaps it against the surface, making a loud, powerful sound that can be heard from a long distance. This behavior is thought to serve different purposes, like communication, disorienting prey, and displaying dominance, especially among males during mating season. Humpback whales may also use tail-slapping to express frustration or warn other group members about danger. In one encounter off the coast of Hawaii, researchers observed a humpback repeatedly slapping its tail after a close encounter with a group of orcas, possibly as a defensive action. This behavior demonstrates the complexities of humpback whale interactions and their ability to communicate and protect themselves with their habitat.

Humpback whales are known not only for their physical displays but also for their incredible vocalizations, particularly their songs. These complex, haunting melodies are sung primarily by males during the breeding season and can last for hours, echoing through the ocean depths. Each song comprises a series of phrases that repeat in a specific pattern, and it's fascinating to point out that all males in a given population sing the same song, which adjusts over time. Scientists believe these songs serve as a form of communication, perhaps to attract females or establish dominance among competing males. In a fascinating case off the coast of Australia, researchers observed a population of humpback whales that had picked a new song pattern from a neighboring group, demonstrating the whales' ability to learn and adapt to new vocalizations. The scenario demonstrates the transfer of culture among humpback whales, indicating that their songs are not instinctive but passed down and improved over generations.

In addition to their captivating songs, humpback whales engage in another fascinating behavior called "spy hopping." This happens when a whale pokes its head vertically out of the water to observe its surroundings above the surface. Spyhopping is commonly observed during social interactions or when whales are curious about boats or other objects in their environment. This behavior gives humans a rare glimpse into these whales' intelligence and curiosity. A notable example of spy hopping occurred in Monterey Bay, California, when a group of humpback whales was seen repeatedly rising out of the water to inspect a nearby whale-watching ship. The whales appeared to be as fascinated by the humans as the humans were by them, which led to a unique interaction of two entirely different species. This behavior indicates that humpback whales are aware of their surroundings and can also show interest in and interact with other species in their environment.

From their mesmerizing songs to their cooperative hunting strategies, these ocean giants demonstrate a level of complexity that continues to captivate scientists and the public alike. Every breach, every song, and every migration tells a story of resilience, adaptation, and the enduring beauty of the humpback whale.

7

The Whale Shark

The whale shark, scientifically known as *Rhincodon typus*, is the ocean's largest fish, known for its massive size and whale-looking appearance. Its size may be intimidating—up to 40 feet or more— but this gentle giant is known for its gentle demeanor and slow movements through water. Unlike the fearsome predators that the name "shark" is usually attached to, the whale shark is a filter feeder that eats mainly plankton, tiny fish, and other small marine organisms. It glides through the ocean with its mouth open, filtering massive amounts of water to extract its food. The whale shark can be found in warm, tropical waters worldwide, from the coast of Australia to the Gulf of Mexico, attracting divers and marine biologists looking to see this spectacular, graceful giant up close.

Whale sharks stand out not only by their size but also by their coloration and patterns. Their skin, up to 4 inches thick, is a beautiful combination of pale yellow or white spots and stripes on a dark gray or blue background. These patterns are not only visually stunning but also are important for studying and protecting these creatures. Each whale shark has a unique design with spots and stripes on its back that are just as unique as a human fingerprint. These patterns help researchers to identify individual whale sharks, track their movements, and collect valuable information about population sizes and migration routes. In recent years, technology has become important in this research.

Conservation organizations have created databases where divers and scientists can upload photos of whale sharks, which are then identified using pattern recognition software. This has improved the way scientists study whale sharks, making it easier to track their movements and understand their behavior around the world. The use of technology in this way shows the importance of innovative approaches to wildlife conservation, particularly for species as difficult to track as the whale shark. Another example is the "Wildbook for Whale Sharks" project, which collects photographs submitted by divers around the world into a database of whale sharks. This citizen science initiative has greatly improved the understanding of these elusive animals, even leading to the discovery of new whale shark gathering sites.

The social habits of whale sharks are another fascinating aspect of their lives. While they are usually solitary creatures, they have been known to gather in large numbers in areas with plenty of food, especially during seasonal plankton blooms. One of these gatherings takes place in the waters off the coast of the Philippines, where hundreds of whale sharks converge each year to feed. This event, known as a "whale shark aggregation," gives researchers a rare opportunity to study these animals up close and collect data on their health, behavior, and interactions. Despite their size, whale sharks are surprisingly gentle and non-aggressive during these aggregations, usually swimming peacefully alongside one another and sharing available food sources. These gatherings also highlight the importance of certain habitats for whale sharks' survival, as they rely on these abundant feeding grounds to sustain their massive bodies.

Whale sharks are also impressive for their strange and uncommon migration patterns. These ocean giants are known to travel thousands of miles worldwide, feeding grounds usually separated by great distances. Yet, many aspects of their migrations remain unknown. For example, one tagged whale shark traveled over 8,000 miles in just over two years, from the Gulf of California to the Marshall Islands, an incredible journey that has puzzled scientists. This type of long-distance migration is thought to be due to a desire for food and suitable breeding grounds, but the exact causes and strategies underlying these movements remain unknown. This mystery only adds to the whale shark's attractiveness as new discoveries about their behavior reveal how little we know about these gentle giants.

Another factor influencing whale sharks' survival is their ability to adapt to their surroundings. While whale sharks are best known for their tropical and subtropical ocean areas, they have been seen in other environments, like the warm waters off the coast of the Seychelles and cooler waters near Japan. Their ability to travel long distances in search of food is assisted by their slow, intentional swimming style, which saves energy and allows them to cover long distances in the ocean.

In 2011, researchers tagged a whale shark off the coast of Panama. They followed it across the Pacific Ocean to the Philippines, covering over 12,000 miles in three years. The incredible journey showed the whale shark's impressive migratory abilities and the importance of international cooperation in conservation efforts, as these creatures travel through multiple countries' waters. Understanding and protecting their migratory routes is important for ensuring the species' survival, as they face constant challenges from fishing, shipping, and climate change.

Whale sharks and humans have a complicated relationship that includes appreciation and exploitation. On the one hand, whale sharks have become a symbol of marine conservation efforts, with their images used to raise awareness about the importance of protecting our oceans. Ecotourism focused on whale sharks has been successful in areas where these gentle giants are known to gather, like

Mexico's Isla Holbox and the Philippines' Donsol Bay. Tourists travel to these places to swim alongside the world's largest fish, bringing valuable economic benefits to local communities. Even so, this interaction comes with the task of ensuring that the animals are not harmed. Strict guidelines have been implemented in many areas to prevent overcrowding, limit the number of boats, and ensure swimmers stay safe from the sharks. These steps are critical for balancing the benefits of ecotourism with the need to protect whale sharks and their habitats.

Unfortunately, not every human interaction with whale sharks is good. Despite being protected in many countries, whale sharks face threats from illegal fishing and accidental captures. In some areas, their fins, meat, and oil are in high demand, leading to the capture and slaughter of these magnificent creatures. The demand for their fins, in particular, has led to a black market that risks whale sharks and many other shark species.

Plus, because whale sharks frequently swim near the surface, they are at risk of collisions with ships, which can result in serious injuries or even death. The story of one whale shark named "Grandpa" by researchers in Taiwan shows the danger. Grandpa was a well-known whale shark who had been tracked for several years before his tragic death, which was most likely caused by a collision with a large vessel. This incident highlights the ongoing threats to whale sharks and the urgent need for more effective conservation measures.

The reproduction of whale sharks remains one of the most mysterious aspects of their existence, owing to a lack of observations of their breeding behavior in the wild. Whale sharks are ovoviviparous, meaning the female keeps fertilized eggs inside her body until they hatch before the young sharks are born alive. In 1995, a pregnant whale shark was caught off the coast of Taiwan, giving researchers a rare glimpse into the process. The shark was discovered to carry over 300 pups at various stages of development, making it one of the largest known litters in any vertebrate. This discovery was groundbreaking, revealing that whale sharks give birth to many young at once, perhaps compensating for the high mortality rate of newborns in the ocean. However, the mystery of where these young sharks are born and how they grow to adulthood remains largely unsolved, with scientists still looking for the whale shark nurseries.

Whale sharks also have an extraordinary sense of direction, which is thought to be due to their highly developed sensory organs. A particular organ, the *Lorenzini ampullae*, allows them to sense the electric fields generated by other living creatures, allowing them to locate prey even in the ocean's darkest depths. In addition, whale sharks have a keen sense of smell, which they use to detect scents in the water and guide them to feeding grounds.

These adaptations are critical for their survival, given the large and often crowded habitats in which they live. A notable example of their incredible navigating abilities is the story of "Wanda," a whale shark tagged in the Gulf of Mexico and later tracked as she crossed the Atlantic Ocean to reach the waters off the coast of Africa. Wanda's journey, which took several months and covered thousands of miles, showed whale sharks' incredible navigational skills.

Whale sharks' unique anatomy also helps them succeed as filter feeders. Their mouths, which can be as wide as five feet, are lined with up to 300 rows of tiny teeth. Despite having teeth, whale sharks do not chew or tear their food. Instead, these teeth are thought to help guide food to the gill rakers, which trap and filter nutrients from the water. Furthermore, whale sharks have a highly developed olfactory system that allows them to detect plankton and other food sources from long distances. This keen sense of smell is especially important in the ocean, where food is naturally scattered. Local fishermen in the Maldives often observe whale sharks appearing seemingly out of nowhere when the waters become rich in plankton, showcasing these creatures' incredible sensory abilities.

Despite the challenges, there are reasons to be hopeful about the future of whale sharks. International efforts to protect whale sharks have gained momentum, with several countries approving sanctuaries and enforcing stricter fishing regulations. These stories of change and collaboration demonstrate that these magnificent ocean giants can have a brighter future with the right combination of education, enforcement, and community involvement.

8

The Giant Squid

The giant squid, *Architeuthis dux*, is one of the deep sea's most elusive creatures. These enormous cephalopods, which can reach lengths of 43 feet for females and 33 feet for males, are among the world's largest invertebrates. Their body structure is intriguing and intimidating, with eight arms and two longer tentacles lined with suction cups equipped with sharp, serrated rings. These adaptations help the squid to grab and hold onto prey with incredible strength, which is needed in the dark depths of the ocean where they hunt.

Another thing that distinguishes the giant squid is its massive eyes, the second largest in the animal kingdom, measuring up to 10 inches in diameter. These massive eyes are perfectly adapted to detect even the smallest traces of light in the deep ocean, allowing the squid to see predators like sperm whales from afar. In 2012, researchers from Japan's National Museum of Nature and Science captured the first footage of a giant squid in its natural habitat, revealing the creepy glow of its eyes as it swam gracefully into the abyss. The sighting showed priceless details about how these creatures move around the inky blackness of the deep sea.

The giant squid's anatomy is also designed to withstand the high pressures of the deep ocean. Giant squids do not have a hard skeleton. Instead, their bodies are made of soft, gelatinous tissue that allows them to cope with the crushing pressures of the deep sea, where they typically live at depths ranging from 1,000 to 2,000 feet. This ability to survive in these harsh conditions has made the giant squid a source of great attraction and mystery, leading to many stories and myths. A particular story from the nineteenth century tells of a giant squid that allegedly attacked a French naval ship, leading to tales of sea monsters that terrorized sailors and inspired imaginations for generations.

The giant squid's deep-sea adaptations are its physical form, behavior, and survival strategies. In the darkness of the ocean depths, where sunlight cannot reach, the giant squid also relies on its heightened senses and unique biological features to navigate, hunt, and avoid predators. One of the aspects of its behavior is the use of bioluminescence. While the giant squid is not bioluminescent, it does live in an environment where many of its prey and predators, including lanternfish and other deep-sea creatures, emit light. The squid's large eyes are highly sensitive to bioluminescent lights, allowing it to detect and track even the smallest signs of light, which usually signal the presence of prey or predators.

A recorded encounter between a giant squid and a sperm whale demonstrates the cephalopod's formidable defense mechanisms. In one incident, a live giant squid was seen lashing out at a sperm whale with its long tentacles, apparently trying to wrap around the whale's head to avoid the attack. Although the whale eventually won that battle, this encounter demonstrates the giant squid's function as predator and prey in the deep-sea ecosystem.

The giant squid's ability to grow to massive sizes in the deep ocean also raises curiosity about its growth patterns and life cycle. Unlike many shallow-water marine creatures, faced with predators and competitors at all stages of life, the giant squid's deep-sea habitat is relatively predator-free, allowing it to grow. This growth is made possible by a diet mainly consisting of deep-sea fish and other squid species, which it captures using its tentacles and strong beak. Results like these add to our understanding of the giant squid's role in the marine food chain and how it adapts to the mysterious, largely unexplored deep-sea habitat.

A fascinating part of the giant squid's life cycle is the mystery of where it spawns. Unlike many fish, which return to certain spawning grounds, the giant squid's deep-sea habitat makes it difficult to pinpoint exact locations where they may lay their eggs.

However, scientists have gained little knowledge by analyzing specimens washed ashore or caught by deep-sea fishermen. Male giant squids are believed to transfer sperm packets to the female through a specially adapted arm known as the hectocotylus. These sperm packets can then be stored for a long time, allowing the female to fertilize her eggs when the conditions are good.

Some scientists also believe that giant squids may spawn at very deep levels, where their eggs and the young can develop in relatively secure environments, out of reach of most predators. The discovery of small, juvenile giant squids in deep-sea searches supports this theory. However, much about their early stages remains unknown. Because of the extreme conditions in their habitat, their growth is likely to be slow, with juveniles potentially spending years growing in the dark depths before reaching their massive adult size.

The modern era of deep-sea exploration has provided rare but valuable opportunities to study giant squids in their natural habitats. One of the most significant breakthroughs came in 2004 when a team of Japanese scientists led by Tsunemi Kubodera took the first photographs of a live giant squid in its natural environment. The squid was discovered about 900 meters (3,000 feet) off the Ogasawara Islands, southeast of Japan. The images showed a 26-foot-long squid thrashing its long tentacles to free itself from a baited line. This event marked an important turning point in marine biology, providing the first true peek into the life of this elusive creature.

Building on their previous success, the same team achieved another accomplishment in 2012 when they captured the first video footage of a giant squid in the deep sea. The footage, captured at approximately 630 meters (2,067 feet), revealed the squid's remarkable agility and movement as it journeyed the deep, dark waters. The video provided scientists with valuable information about the squid's behavior, including its hunting techniques and use of bioluminescence. However, the footage raised new questions about the species' distribution, feeding habits, and life cycle. While these observations have increased our understanding, much of the giant squid's life remains unknown, which continues to captivate and challenge marine scientists.

Despite advances in technology and science, the giant squid remains one of the ocean's most mysterious creatures. It has captivated sailors, scientists, and storytellers alike for centuries, leading to a far-reaching cultural impact outside of science. This mythical creature has become a symbol of the unknown, appearing continuously in pop culture. From books to movies, the giant squid has been portrayed as a terrifying predator and a misunderstood concern of the deep sea. In Jules Verne's classic novel "Twenty Thousand Leagues

Under the Sea," the submarine Nautilus encounters a monstrous squid, leading to a thrilling battle between man and beast. This representation established the giant squid's reputation as a fearsome creature. In recent years, films like Pirates of the Caribbean: Dead Man's Chest has also drawn inspiration from the Kraken legend, popularizing the giant squid as an image of mystery and the supernatural. While typically exaggerated, these cultural depictions point out humanity's strong interest in the giant squid and its place in the hierarchy of underwater mythology.

As researchers continue to explore the deep sea, new discoveries about the giant squid will likely emerge, providing more details about its life cycle and the mysterious world it dwells in. Its ability to survive in one of the most hostile environments on Earth, combined with its uncommon appearances and strong adaptations, ensures it will remain one of the ocean's barely known giants. Whether as a mythological figure, a scientific subject, or a cultural icon, this creature will undoubtedly captivate the world for generations to come.

9
The Great White Shark

The great white shark, *Carcharodon carcharias*, is considered the ocean's deadliest predator. This apex predator's reputation is well-deserved. It has a specially designed anatomy that allows it to hunt precisely and brutally. The great white shark's powerful, torpedo-shaped body allows it to reach up to 25 mph speeds, making it one of the fastest sharks in the ocean. This speed, combined with its size — reaching lengths of up to 20 feet and weighing up to 5,000 pounds—gives the great white a strong presence in the marine environment. The shark's body is designed for maximal effectiveness in water, with an aerodynamic design to reduce drag and a crescent-shaped tail for powerful propelling power.

The great white shark's teeth are one of its trademark features. Each tooth is triangular, serrated, and extremely sharp, allowing it to easily tear through its prey's flesh. The great white has multiple rows of teeth; when one is lost or damaged, another quickly replaces it within 24 hours. This continual regeneration of teeth keeps the shark prepared for its next meal. The great white's jaws are also perfectly suited for hunting, with a bite force of more than 4,000 pounds per square inch—one of the strongest bites in the animal kingdom. The great white's incredible bite force and razor-sharp teeth allow it to hunt large prey such as seals, sea lions, and even smaller whales.

Apart from its physical abilities, the great white shark's senses have been specially programmed to detect prey at long distances. One of the most remarkable adaptations is its ability to detect electromagnetic fields emitted by other animals. This works by using a unique system of gel-filled pores. These *Lorenzini ampullae* are positioned around the shark's nose, allowing it to detect small electrical impulses generated by the muscles and heartbeats of its prey, even when the prey is hidden or camouflaged. This ability gives the shark an advantage when locating prey in the ocean's massive and dark depths.

Additionally, the great white has an acute sense of smell, allowing it to detect a single drop of blood in an Olympic-sized swimming pool (about 165 feet long). This heightened sense of smell is particularly useful in the ocean, where the ability to track a scent trail over long distances can mean the difference between survival and death. Their olfactory bulbs are among the largest in the animal kingdom, making them an extremely successful predator capable of detecting injured or distressed prey from long distances. Great white sharks have been seen converging on the carcasses of large marine mammals off Guadalupe Island, drawn by the scent of blood from miles away.

The great white's eyes also play an important role in its hunting strategy. Unlike the popular belief that sharks have poor eyesight, the great white has excellent vision, especially in low-light conditions. This is due to the *tapetum lucidum*, a reflective layer of cells behind the retina that improves vision in the ocean's dark depths. This ability is useful at sunrise and sunset when the shark and its preferred prey - seals are most active. An example of this is the behavior of great white sharks in the Farallon Islands near San Francisco, where they have been observed attacking seals in the early morning hours when visibility is low, using their vision and other sensory abilities.

The great white shark's hunting abilities are equally impressive as its physical attributes. One of the most common strategies is the "ambush" attack, where the shark approaches its prey from below using camouflage and stealth. Its white underbelly blends into the bright surface of the water, making it nearly invisible to prey looking down from above. Once within striking range, the shark accelerates upward with shocking speed and power, often breaching the surface in a wonderful display of force. This method is more effective against fast prey like seals, which the great white often prefers to feed on, off South Africa's Seal Island.

This breaching behavior demonstrates the shark's physical strength and its strategic approach to hunting. The shark's conical snout and powerful jaws, lined with serrated teeth, are perfectly designed to grip and tear into its prey's flesh, reducing the possibility of escape. After its first strike, the shark usually circles back to eat its prey, saving energy and minimizing the risk of injury. This hunting method has been observed in a number of shark hotspots around the world, including Australia's Neptune Islands, where researchers have studied the great white's precise and calculated predatory behavior.

The great white's hunting abilities extend outside ambush techniques. In some areas, these sharks have been observed participating in social hunting, where more than one shark targets a single prey item, like a whale carcass. This cooperative behavior, though very rare, shows the great white's ability to adjust its hunting strategy based on the availability of prey. In the seas of New Zealand, great white sharks have been observed feeding on sperm whale carcasses, with several sharks taking turns biting and feeding, indicating a level of social coordination new to these solitary hunters.

Occasionally, they can be seen in partially organized groups, particularly near rich feeding grounds. These groups are usually temporary, forming and dissolving as sharks follow seasonal migrations of their prey. A fascinating example of this social behavior can be seen off the coast of South Africa, near Seal Island, where great white sharks come together in large numbers during the winter to hunt seals. In these cases, the sharks show very rare tolerance for one another, sometimes even sharing small prey.

A social hierarchy also appears in these gatherings, with bigger, more dominant sharks usually having supremacy over smaller ones. This ranking is established through displays of power rather than direct aggression. A dominant shark may circle or bump a rival to assert its dominance, as observed by researchers studying these creatures in locations like the Neptune Islands off Australia. These interactions suggest that great white sharks can engage in complex social behaviors besides predation, pointing to intelligence and adaptability, contradicting their reputation as just mindless killers.

Great white sharks have also been observed engaging in what can be described as playful behavior. Sharks have been seen investigating wreckage floating off Guadalupe Island, Mexico, out of curiosity rather than hunger. They may nudge or bite at objects, almost like they are inspecting their surroundings. This behavior is interesting because it suggests that great white sharks have the ability to explore and learn. Such observations call into question the conventional perception of these sharks as purely primitive predators, opening new avenues for studying their cognitive capabilities and behavioral complexities.

As researchers figure out the mysteries of their behavior, it becomes clear that great white sharks are more than just mindless killing machines; they can engage in complex social interactions and apply adaptive strategies. The study of great white shark behavior and social structures will continue, and each new discovery will challenge our understanding of these apex predators. Protecting great white sharks ensures the preservation of their tricky social systems and how important they are in maintaining marine ecosystem balance.

10

The Leatherback Sea Turtle

The leatherback sea turtle, *Dermochelys coriacea*, is one of the oldest species in our oceans, with a lineage dating back more than 65 million years. Unlike other sea turtles, the leatherback has survived multiple mass extinctions, showing its amazing adaptability and tenacity. The leatherback's history may be traced back to when dinosaurs roamed the Earth, making it a living relic of a prehistoric age. The fact that it survived during these extinction events is proof of its unique biological features, which have allowed it to survive in a wide range of marine environments.

Leatherback sea turtles hold the record of both the largest and fastest sea turtles in the world. These ancient creatures can grow to lengths of up to 7 feet (over 2 meters) and weigh between 600 to 1,500 pounds, with some reaching even 10 feet in length. Known for their incredible speed in the water, leatherbacks can swim as fast as 22 miles per hour, making them highly efficient ocean travelers. Their unique adaptations and their streamlined shape and powerful front flippers, help them to glide through the water with ease and cover vast distances across the oceans.

The leatherback's ability to dive to deep depths is also due to its unique shell construction and exceptional cardiovascular adaptations. During deep dives, the turtle's heart rate reduces drastically, allowing it to stay underwater for up to 85 minutes while saving oxygen. This physiological characteristic allows leatherbacks to make deep dives throughout the day for their main meal, jellyfish. Scientists tracking leatherback behavior have reported dives lasting more than an hour at depths below 4,000 feet, showing the turtle's extraordinary adaptation to the difficulties of deep-sea survival.

One example showing the leatherback's deep-diving prowess happened in 2015 when a team of researchers tracked a tagged leatherback turtle named "Darla" off the coast of New Zealand. Darla surprised scientists by diving to a depth of more than 4,200 feet, equivalent to the height of 14 Statues of Liberty placed on top of one another. This trip into the deep ocean revealed knowledge about leatherbacks' feeding habits and migration patterns, which helped to support conservation efforts to protect these ancient marine creatures.

In addition to its deep-diving ability, the leatherback has one of the lengthiest migratory patterns among aquatic creatures. Leatherbacks embark on transoceanic journeys that cover thousands of miles and cross oceanic regions. The migrations are caused by the seasonal availability of jellyfish, which are frequently found in nutrient-rich waters. Leatherbacks nesting on Costa Rican beaches have been reported to travel as far as the South Pacific near New Zealand, almost 7,000 miles. This broad migratory range shows the leatherback's endurance and navigational abilities.

The leatherback sea turtle's migrating and nesting routines need endurance and perfectly calibrated responses to environmental stimulation. These turtles use the Earth's magnetic fields to travel huge distances in the ocean, a phenomenon that scientists are only beginning to understand. Based on research by the University of North Carolina, leatherbacks can detect very little differences in magnetic fields, allowing them to identify certain nesting beaches with surprising precision. This magnetic sensitivity and heightened sense of smell also guides them on the dangerous journey back to their birthplaces. In 2018, a leatherback turtle was tracked as it maneuvered through the Pacific's major shipping lanes, showcasing its ability to avoid threats while staying on course for thousands of kilometers.

Another incredible part of leatherback nesting is the ability to return to the same beach year after year, usually with pinpoint accuracy. This habit, also known as "site fidelity", has been seen in leatherbacks breeding along Florida's eastern coast, where they return to the same few miles of beach yearly. Although site fidelity can be a two-edged sword. If or when a beach becomes unsafe because of erosion, pollution, or human activity, leatherback turtles may still return regardless of the poor conditions. This has been observed in some regions of Malaysia, where once-pristine nesting habitats have been degraded by beachside construction. Efforts to help solve these problems include creating marine protected zones and collaborating with local communities to regulate beach lighting, which may mislead both nesting females and hatchlings and threaten the survival of these ancient marine creatures.

The leatherback's survival ability for more than 65 million years shows its evolutionary success. Yet, modern challenges bring about new concerns that even this ancient species must face. One of the most severe issues is plastic pollution in the oceans. Leatherbacks mostly eat jellyfish and tend to mistake floating plastic bags for their natural prey. Consuming plastic can cause digestive problems, malnutrition, and even death. A devastating incident happened in 2016 when a leatherback washed up in Florida with nearly 13 pounds of plastic trash in its stomach. This terrible tragedy attracted global attention to the harmful effects of plastic pollution on marine life, stating the serious need to reduce plastic waste. Another major challenge facing leatherback sea turtles is the detrimental effects of climate change on their nesting locations, particularly the effect it has on hatchling sex ratios. Leatherbacks often nest on tropical and subtropical beaches, where the temperature of the sand influences the gender of the hatchlings through a process known as **temperature-dependent sex determination**. Warmer sands usually produce more female hatchlings, while cooler sands produce more males. This natural balance is necessary for maintaining healthy populations.

However, as global temperatures rise, there is rising concern about the leatherbacks' long-term survival due to uneven sex ratios. Nesting sites produce a higher percentage of female hatchlings, potentially leading to a male shortage in the population over time. Researchers in Costa Rica, one of the main nesting sites for leatherbacks, have already noticed a major spike in female hatchlings in recent years, a trend that could destabilize the species' population if it continues unchecked.

This shift in sex ratios poses a serious threat to leatherbacks, as a lack of males may disrupt mating patterns and reduce the species' ability to reproduce effectively. To address this, conservationists have implemented some proactive strategies to reduce the impact of rising temperatures on nesting sites. In

Costa Rica, efforts have been made to shade nests with natural materials such as palm fronds, which can lower sand temperature, or to relocate nests to cooler areas where the temperature is better suited to producing a balanced sex ratio. These measures are important for ensuring that leatherback populations survive despite the rising temperatures.

Despite these challenges, conservation efforts have effectively safeguarded leatherback sea turtles. Community-led initiatives have successfully reduced poaching while protecting nesting beaches in some areas. In Gabon, for example, local conservation groups collaborated with international organizations to monitor and protect leatherback nests, significantly increasing hatchling survival rates. In Trinidad, where leatherbacks face many challenges during the nesting season, like coastal construction and illegal poaching, leatherbacks' unwavering return to the beaches of Grande Riviere has sparked a community-led conservation effort. Locals, who once saw turtles as a source of meat, now fiercely protect them, organizing nightly patrols during the nesting season to ensure the turtles and their nests are not disturbed.

In some areas, leatherback sea turtles have been observed changing their nesting habits in response to environmental changes. In the Pacific, for example, rising sea levels and increased storm activity have seriously damaged some nesting beaches.

Leatherbacks have responded by laying their eggs on beaches farther inside or at higher elevations, demonstrating behavioral flexibility. While this adaptability provides a bit of optimism, it also shows the importance of addressing the larger environmental issues that threaten the turtles' survival. If their nesting sites deteriorate, these adaptive strategies may be too small to sustain the population.

Despite their remarkable resilience, leatherback sea turtles now face a precarious future, with populations rapidly declining worldwide. Current estimates suggest that, without effective conservation efforts, these ancient giants could go extinct within the next 15 to 20 years. As survivors of Earth's most devastating extinctions, these turtles have endured countless threats over millions of years. Now, with time running out, one critical question remains: will they manage to survive this final challenge?

11

The Manta Ray

The manta ray, *Manta birostris*, is another of the ocean's most iconic and majestic creatures, recognized by large, wing-like pectoral fins measuring up to 23 feet across. Unlike other rays, mantas don't have a stinger on their tail, making them completely harmless to humans despite their enormous size. The manta's body is flat and long, with a wide, triangular shape allowing for graceful gliding movements in the ocean - similar to a bird flying in the sky. This unique anatomy, especially their wide, flat bodies, is not just for show; it is important to their survival, which allows them to swim and feed efficiently.

One of the most captivating aspects of manta ray anatomy is their cephalic fins, which are positioned on both sides of their mouths. Manta rays are excellent filter feeders because they can spread their fins and funnel plankton and small fish into their mouths while swimming. They swim in slow loops to take in as much food as possible, making this feeding method effective and almost effortless. This feeding method can be seen in large groups, with dozens of mantas swimming in perfect sync, each following the next. These impressive feeding "trains" demonstrate these creatures' complex social behaviors.

Manta rays are migratory, traveling far across oceans to find food, mate, and reproduce. In the oceans of the Maldives, where large groups gather to feed on plankton during the monsoon season. These gatherings, known as "feeding aggregations," are incredible, with dozens or even hundreds of manta rays flying through the water in search of food. The reasons for their migration patterns are still unknown. Still, scientists believe they are influenced by seasonal changes in water temperature and food availability. Tracking manta ray movements using satellite tagging has also revealed more information about their behavior, helping researchers to better understand and protect these endangered species.

Apart from their physical appearance, manta rays have a high level of intelligence that separates them from many other aquatic creatures. Researchers discovered that manta rays have the largest brain-to-body ratio of any cold-blooded fish - 5:1, similar to humans, and their brains are even larger than many mammals. This intelligence is reflected in their complex social behaviors, such as recognizing themselves in a mirror, previously thought to be confined to the most intelligent animals like dolphins, elephants, and the great apes. This ability to recognize themselves suggests a rare level of self-awareness in the animal kingdom.

Manta rays also have complex and unique communication methods. These rays are frequently seen in groups, especially during feeding, mating, and cleaning activities. One interesting aspect of their social behavior is how they interact at cleaning stations, where they assemble in large numbers to have parasites and dead skin removed by the cleaner fish. These cleaning stations, typically found near coral reefs, turn into busy areas of activity, with manta rays queuing up and taking turns like they have some protocol. The social structure of these gatherings can also be quite complex, with larger and more dominant rays getting higher priority at the cleaning stations, showing a hierarchy within these groups.

Manta rays' intelligence can also be seen in their interactions with humans, especially in areas like the Great Barrier Reef and the Maldives, where these creatures occasionally approach divers out of curiosity. Unlike many other large marine animals, manta rays are known for their peaceful nature and seem to continually seek human interaction. Many divers have shared their close encounters with manta rays, describing their peaceful and calm behavior, even allowing themselves to be gently touched.

A manta ray named "Freckles" approached a group of divers in Hawaii, seeking help removing fishing hooks attached to her wing. The ray's deliberate approach and patience as the divers tried to free her showed intelligence and a rare level of trust and communication, pointing out the bonds these creatures can form with humans. Another diver described how a manta ray swam so close to her that its wing brushed against her. A seemingly deliberate and gentle touch left her in awe of the creature's intelligence and awareness. These encounters indicate manta rays' distinct nature and their frequently strong impact on those lucky to share the water with them.

Manta rays' graceful movements are often described as "flying" through the water, making them an interesting sight for divers and marine enthusiasts. These gentle giants glide effortlessly, propelled by their large pectoral fins with a fluidity that appears almost supernatural. Manta rays rely on clean, healthy oceans for survival because they are filter feeders that consume a lot of plankton. Their feeding behavior, frequently observed in rich, nutritious waters, shows the importance of protecting these marine environments. Manta rays' health is essential to the health of our oceans, making them an important species in conservation efforts.

Manta rays' intelligence and gentle nature have earned them respect in many cultures. In Polynesian mythology, mantas are depicted as protective spirits who guide and watch over fishermen. In some traditions, they are considered reincarnations of ancestors, demonstrating the cultures' deep admiration and love for the animal. A perfect example comes from Micronesia's Yap Island; manta rays are considered sacred here, and the locals have a long history of coexisting with them. The island's people believe that mantas rays are the spirits of their ancestors, leading to profound respect and protection for the species. The community-led conservation efforts on Yap have been very successful, resulting in a flourishing manta ray population and an ecotourism industry that benefits both locals and the environment.

However, the characteristics that make manta rays attractive and approachable also put them at risk. Manta ray populations are declining in some regions because of their slow reproductive rate and increased threats from pollution, climate change, and illegal fishing. The demand for manta ray gill plates in traditional medicine, particularly in parts of Asia, has led to excessive fishing practices, endangering the lives of these majestic creatures.

In 2014, Indonesia declared its waters a sanctuary for manta rays, banning their capture and creating one of the world's largest manta ray sanctuaries. This decision has been commended as an important step forward in global manta ray conservation efforts.

12

The Greenland Shark

The Greenland shark, often called the "ancient mariner of the deep," and with good reason, is among the oldest living vertebrates on the planet. This exceptional longevity demonstrates the species' resilience and provides a rare glimpse into the history of the oceans. Living in the freezing temperatures of the North Atlantic and Arctic Oceans, the Greenland shark's slow metabolism and deep-sea habitat have contributed to its long life. This species' ability to survive in this harsh environment over a long period is a wonder, raising interesting questions about the secrets of its long lifespan and adaptation to extreme conditions.

One of the most incredible aspects of the Greenland shark's life is its slow growth rate. These sharks grow at a rate of about one centimeter per year, so a five-meter-long shark could be centuries old. This slow growth and late maturity—many do not reach reproductive age until they are over 100—severely affects the species' survival. The Greenland shark's slow development risks it from threats like overfishing and climate change, as the populations take a long time to recover from declines. Despite this, these sharks have survived for centuries, adapting to the icy dark depths where only a few large predators can exist.

In addition to its slow metabolism, the Greenland shark has unique physical traits, allowing it to withstand the deep ocean's high pressures and cold temperatures. Its blood contains a high concentration of trimethylamine N-oxide (TMAO). This chemical defends proteins and cell membranes from the effects of pressure and cold. This trait helps the Greenland shark maintain proper bodily functions at depths greater than 2000 meters (more than 6,560 feet). This allows it to migrate vertically within the water column in search of prey. Traveling such a wide range of depths is useful in the Arctic, where food sources are scattered across wide regions.

One of the Greenland shark's major adaptations is its slow metabolism, which is essential for survival in the cold, low-nutrient waters of the Arctic. Unlike many other shark species that live in warmer climates, the Greenland shark has a slow metabolism that allows it to live for centuries, with some individuals estimated to be over 500 years old. This slow metabolic rate conserves energy in an environment where food is scarce and contributes to the shark's life span.

Scientists and marine biologists are astonished by the Greenland shark's extraordinary lifespan, a living window into our oceans' ancient past.

Despite its sluggish nature, the Greenland shark is a fierce predator capable of hunting many kinds of prey in its icy environment. As revealed by what was found in their stomachs, these sharks are known to feed on a number of aquatic species, like fish, seals, and carcasses. Their slow, stealthy movements and ability to remain undetected in the dark waters also give them an advantage as ambush predators, silently creeping up on their prey.

Inuit hunters discovered a remnant of an eaten polar bear carcass washed ashore. Initially, they wondered what could have attacked such a powerful predator. Further investigation revealed that the wounds matched the bite marks of a Greenland shark. This rare incident demonstrated the shark's ability to take on even the Arctic's top predators, although through scavenging rather than active hunting. These events indicate the Greenland shark's ability to survive by seizing every opportunity.

Interestingly, there have also been reports of Greenland sharks preying on seals while they sleep, showing the shark's opportunistic hunting strategy. Unlike faster sharks that hunt aggressively, the Greenland shark hunts patiently and stealthily. This has made some researchers nickname the Greenland shark a "sleeping giant," one that waits for the right opportunity before striking. This method of attack shows not only the shark's adaptability but also its capacity to conserve energy in the cold environment in which it lives.

The Greenland shark's slow movement has led it to scavenge on carrion, including the remains of large marine mammals such as seals and whales that sink to the ocean floor. This scavenging behavior proves beneficial in the Arctic, where a scarcity of prey forces the shark to make the most of every available food source. There have even been reports of Greenland sharks found with polar bears, and reindeer remains in their stomachs, demonstrating their ability to consume a wide range of prey and make the most of whatever the environment offers.

A fascinating story demonstrating the Greenland shark's resilience and adaptability is a 2013 study in which researchers attached cameras to seals' backs in Svalbard, Norway, to study their behavior under ice. To their surprise, the cameras caught footage of Greenland sharks swimming slowly beneath the ice, apparently in areas where seals were known to rest. This discovery clarifies the shark's hunting strategies and ability to exploit the Arctic's unique conditions.

Myth and mystery have surrounded the Greenland shark for years, especially among the Inuit and other indigenous peoples of the Arctic regions. This giant of the deep is feared and admired, inspiring many tales passed down through generations. The myth of

Skalugsuak, a mythical animal created by the sea goddess Sedna and regarded by some Inuit tribes as the first Greenland shark, is one of these stories. The tale claims that Sedna, upset by human recklessness, created Skalugsuak to patrol the cold, dark waters, to ensure that people followed the natural order. In addition to highlighting the Greenland shark's cultural significance, this story illustrates the close relationships between the Arctic people and their environment, where maintaining the delicate balance of life is frequently important for survival.

Adding to the shark's mystique is its association with the concept of eternal life. The Greenland shark is the longest-living animal that science has discovered, with some estimated to be over 400 years old. Their extraordinary lifespan has only strengthened the myth that these sharks are old spirits or messengers of knowledge, watching the world change over generations. Their lifespan raises the question of what mysteries they might keep, having seen time pass in the remote, unexplored Arctic.

In a 2016 study, researchers measured the ages of 28 Greenland sharks using radiocarbon dating, revealing a real-life example of the shark's exceptional survival. They discovered that the oldest shark in the study might be older than the United States of America, having most likely been born between 1504 and 1744. The Greenland shark's significance as a living relic of the ocean was made clear by this astounding discovery, which shocked biologists. The shark's outstanding adaptability to its harsh environment can be seen by its ability to survive through several human generations and to witness changes in the Arctic environment over centuries. These discoveries deepen our understanding of this elusive species and show the importance of protecting such amazing creatures, which are the vital link to the Earth's ancient past.

13

The Future of the Ocean Giants

Imagine a young humpback whale, barely old enough to navigate the ocean, struggling amidst the constant noise of shipping traffic. This isn't an isolated story—it's a shared experience among ocean giants confronting us with the growing threats they face. From whales and sea turtles to sharks and manta rays, these magnificent creatures are grappling with climate change, pollution, and overfishing, all of which continue to strain their habitats and survival.

Rising ocean temperatures are shrinking the cold habitats of species like the blue whale. At the same time, pollution from plastics fills the waters, harming filter feeders like whale sharks and manta rays. Overfishing and bycatch decimate populations of leatherback sea turtles and Greenland sharks, each casualty underscoring how human activity disrupts the delicate balance of marine ecosystems. Even the noise from shipping routes interferes with the communication of species that rely on echolocation and vocalizations, putting stress on their health and behaviors. Fatal collisions with ships further threaten already endangered species, like the North Atlantic right whale.

But there's still hope. Marine Protected Areas, international agreements, and technological innovations are beginning to make a difference, showing us what's possible when science, policy, and community come together. Conservation efforts like the Papahānaumokuākea Marine National Monument and initiatives under the Convention on Migratory Species illustrate the collective impact of global action. The inspiring recovery of humpback whales in the Southern Hemisphere reminds us that we can foster meaningful change with sustained effort.

Yet, the future of these giants hinges on more than policy; it depends on individuals acting together. We can each contribute—by supporting conservation organizations, reducing plastic usage, or spreading awareness about these issues.

As I conclude this book, I am reminded of the impact these ocean giants have had on me, inspiring a lifelong passion for the mysteries and beauty of the sea. Researching and writing about them has strengthened my commitment to their protection, and I hope this journey into their world stirs something within you as well. Together, let us work toward a future where these majestic beings continue to thrive, reminding us of the ocean's enduring wonders and its essential role in our planet's health.

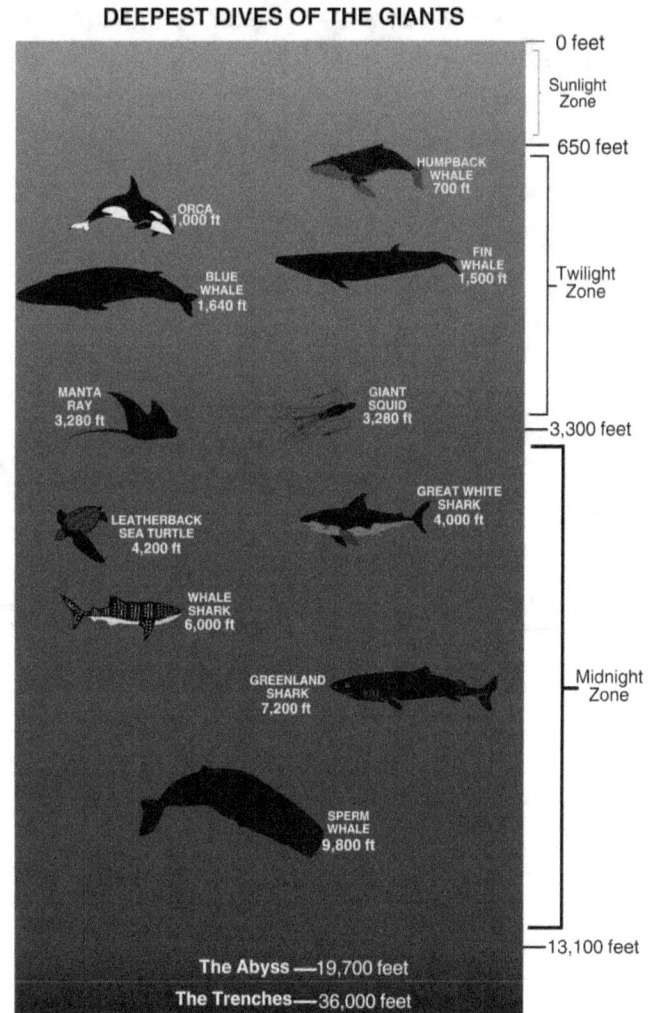

SIZE COMPARISON OF THE OCEAN GIANTS

Glossary

52 Blue: A blue whale that calls at a unique frequency of 52 hertz, often referred to as the "world's loneliest whale," as no other whale has been recorded calling at this frequency.

Abyss: A term often used to describe the deep ocean layers, typically 1,000 to 6,000 meters below the surface, where little to no light penetrates.

Aggregation: A cluster or gathering of things.

Ambush Attack: A surprise attack strategy often involving stealth and speed.

Apex Predator: A predator at the top of a food chain, with no natural predators of its own.

Asymmetrical Jaws: The unique jaw coloration of fin whales, with one side lighter than the other, thought to aid in their hunting techniques.

Baleen Plates: Comb-like structures in the mouths of certain whale species, including blue whales, made of keratin, which are used to filter food from the water.

Bioluminescence: The emission of light by living organisms used for communication, attraction, or camouflage.

Biological Sonar: The method by which sperm whales and some other species detect objects and navigate by producing sound waves that bounce back from objects as echoes.

Blubber: A thick layer of fat beneath a whale's skin, used for energy storage and insulation in cold waters.

Glossary

Bowhead Whale: A long-lived Arctic whale, known for its thick blubber and role in the Arctic ecosystem.

Breaching: The act of a whale or shark leaping out of the water.

Cephalopod: A class of marine animals, including squids and octopuses, characterized by bilateral body symmetry, a prominent head, and arms or tentacles.

Cetaceans: A group of marine mammals that include whales and dolphins.

Citizen Science: Public participation in scientific research, e.g as seen in projects where divers upload whale shark photos to help track individual animals.

Commercial Whaling: The practice of hunting whales for their oil, meat, and other products.

Conservation: Efforts made to protect and restore endangered species and natural habitats.

Conservation Efforts: Actions taken to protect and restore endangered species, like the ban on commercial whaling that aids fin whale population recovery.

Cultural Transmission: The passing down of knowledge and behaviors, such as hunting techniques, within sperm whale pods across generations.

Deep-Sea Adaptation: Specialized physical and behavioral traits that enable organisms like sperm whales to survive and thrive in the extreme conditions of the ocean's depths.

Deep-Sea Habitat: The ocean's deeper zones, often 1,000 to 2,000 feet down or more, characterized by extreme darkness, high pressure, and low temperatures.

Decibels (dB): A unit used to measure the intensity of sound. Blue whale calls can reach up to 188 decibels, louder than a jet engine.

Diving Physiology: The study of the physical adaptations that enable deep-diving animals to survive in low-oxygen and high-pressure environments.

Echolocation: A biological sonar used by animals to navigate and locate objects by emitting sounds and interpreting the returning echoes.

Ecosystem Balance: The state of equilibrium in an ecosystem, maintained by predator-prey relationships.

Electromagnetic Fields: Invisible fields created by electrically charged objects.

Emotional Intelligence: The ability to perceive, understand, and respond to the emotions of others.

Endurance: The capacity for sustaining long-duration activity.

Entanglement: The accidental trapping of whales in fishing gear, which can cause injury or death.

Filter Feeder: An organism that feeds by straining suspended matter and food particles from water, commonly plankton.

Fluke: The tail of a whale, used powerfully to propel the animal through water, especially during long migrations and hunting behaviors.

Giant Squid: A primary prey of sperm whales, known for its large size and elusive nature, often inhabiting deep-sea regions.

Glossary

Gill Rakers: Structures in the mouth that help filter food from water.

Hectocotylus: A specialized arm in male squids used to transfer sperm packets to the female for fertilization.

Ichthyosaurs: Extinct marine reptiles with dolphin-like bodies and powerful tails.

Jörmungandr: A Norse mythological sea serpent that represents the power and mystery of the ocean.

Keystone Species: Species that have a disproportionately large effect on their ecosystems' structure and functioning.

Kraken: A legendary sea monster often associated with giant squids, popular in mythology and literature.

Krill: Tiny shrimp-like creatures that are a primary food source for blue whales, often found in large swarms in nutritious regions of the ocean.

Logging: A behavior where sperm whales float motionless on the ocean's surface to rest, resembling logs drifting in the water.

Lunge Feeding: A feeding strategy used by whales to consume large amounts of water and krill, filtering out the water through baleen plates.

Magnetic Fields: Invisible forces created by the Earth's magnetism, which some animals can detect to aid in navigation.

Marine Protected Areas (MPAs): Ocean regions designated for conservation purposes, where activities like fishing, drilling, and shipping are limited to protect marine life.

Mass Extinction: A period in Earth's history where a large percentage of all species are wiped out, due to drastic environmental changes.

Matriarch: The oldest leader in a group based on experience.

Megalodon: A prehistoric giant shark that could grow up to 60 feet long.

Migration: The long-distance seasonal movement of any creature, typically between feeding and breeding grounds.

Noise Pollution: Disruptive sounds from human activities (e.g., shipping, industrial operations, sonar).

Opportunistic Feeder: An organism, like the fin whale, that adjusts its diet based on food availability in its environment.

Ovoviviparous: A reproductive mode in which eggs develop inside the mother's body and young are born live.

Pakicetids: Prehistoric terrestrial mammals that evolved into modern whales.

Pectoral Fins: The long, narrow fins located on the sides of the whale's body that aid in steering and stability while swimming.

Phytoplankton: Microscopic marine plants that form the foundation of the marine food chain.

Plastic Pollution: Environmental contamination with plastic waste.

Glossary

Pod: A social group of orcas, typically consisting of family members, known for cooperative behaviors and close bonds.

Prehistoric Age: A period dating back to millions of years ago, before human history was recorded, when ancient species roamed Earth.

Quiet Zones: Areas in the ocean where human activity, such as shipping, is restricted to reduce noise pollution and protect marine life.

Ship Strikes: Collisions between whales and ships.

Site Fidelity: The tendency of animals to return to the same location for specific activities.

Social Hierarchy: The organization of individuals within a group based on dominance.

Social Hunting: Cooperative hunting strategies within animal pods, often involving teamwork and role specialization.

Spyhopping: When a whale pokes its head out of the water vertically to observe its surroundings.

Suction Cups: Structures used by squid to grasp prey.

Symbiotic Relationships: Close associations between two or more species, often benefiting both, like the cleaner fish that help remove parasites from sharks.

Throat Pleats: Expandable grooves along the throats of some whales, allowing them to take in large amounts of water and food during feeding.

Underwater Camera: A device used to capture images or videos of marine life.

References

Vasconcellos M, Mackinson S, Sloman K, Pauly D. The stability of trophic mass-balance models of marine ecosystems: a comparative analysis. Ecological modelling. 1997 Dec 1;100(1-3):125-34.

Pyenson ND, Vermeij GJ. The rise of ocean giants: maximum body size in Cenozoic marine mammals as an indicator for productivity in the Pacific and Atlantic Oceans. Biology Letters. 2016 Jul 31;12(7):20160186.

Attard CR, Beheregaray LB, Möller LM. Towards population-level conservation in the critically endangered Antarctic blue whale: the number and distribution of their populations. Scientific reports. 2016 Mar 8;6(1):22291.

Melcon ML, Cummins AJ, Kerosky SM, Roche LK, Wiggins SM, Hildebrand JA. Blue whales respond to anthropogenic noise. PloS one. 2012 Feb 29;7(2):e32681.

Madsen PT, Carder DA, Bedholm K, Ridgway SH. Porpoise clicks from a sperm whale nose—Convergent evolution of 130 kHz pulses in toothed whale sonars?. Bioacoustics. 2005 Jan 1;15(2):195-206.

Cantor M, Shoemaker LG, Cabral RB, Flores CO, Varga M, Whitehead H. Multilevel animal societies can emerge from cultural tr

Goldbogen JA, Pyenson ND, Shadwick RE. Big gulps require high drag for fin whale lunge feeding. Marine Ecology Progress Series. 2007 Nov 8;349:289-301.

Rockwood RC, Calambokidis J, Jahncke J. High mortality of blue, humpback and fin whales from modeling of vessel collisions on the US West Coast suggests population impacts and insufficient protection. PLoS One. 2017 Aug 21;12(8):e0183052.

Coscarella MA, Bellazzi G, Gaffet ML, Berzano M, Degrati M. Technique used by killer whales (Orcinus orca) when hunting for dolphins in Patagonia, Argentina.

Ford JK. Call traditions and dialects of killer whales (Orcinus orca) in British Columbia (Doctoral dissertation, University of British Columbia).

Deecke VB, Barrett-Lennard LG, Spong P, Ford JK. The structure of stereotyped calls reflects kinship and social affiliation in resident killer whales (Orcinus orca). Naturwissenschaften. 2010 May;97:513-8.

Clapham PJ, Leimkuhler E, Gray BK, Mattila DK. Do humpback whales exhibit lateralized behaviour?. Animal Behaviour. 1995 Jul 1;50(1):73-82.

Wiley D, Ware C, Bocconcelli A, Cholewiak D, Friedlaender A, Thompson M, Weinrich M. Underwater components of humpback whale bubble-net feeding behaviour. Behaviour. 2011 Jan 1:575-602.

Rowat D, Brooks KS. A review of the biology, fisheries and conservation of the whale shark Rhincodon typus. Journal of fish biology. 2012 Apr;80(5):1019-56.

Hueter RE, Tyminski JP, de la Parra R. Horizontal movements, migration patterns, and population structure of whale sharks in the Gulf of Mexico and northwestern Caribbean Sea. PLoS One. 2013 Aug 21;8(8):e71883.

Nilsson DE, Warrant EJ, Johnsen S, Hanlon R, Shashar N. A unique advantage for giant eyes in giant squid. Current Biology. 2012 Apr 24;22(8):683-8.

References

Hoving HJ, Roeleveld MA, Lipinski MR, Melo Y. Reproductive system of the giant squid Architeuthis in South African waters. Journal of Zoology. 2004 Feb;264(2):153-69.

Collin SP, Kempster RM, Yopak KE. How elasmobranchs sense their environment. InFish Physiology 2015 Jan 1 (Vol. 34, pp. 19-99). Academic Press.

Martin RA, Hammerschlag N, Collier RS, Fallows C. Predatory behaviour of white sharks (Carcharodon carcharias) at Seal Island, South Africa. JMBA-Journal of the Marine Biological Association of the United Kingdom. 2005 Oct 1;85(5):1121-36.

Lutcavage ME. Human impacts on sea turtle survival. InThe Biology of Sea Turtles, Volume I 2017 Dec 6 (pp. 387-409). CRC press.

Santidrián Tomillo P, Spotila JR. Temperature-dependent sex determination in sea turtles in the context of climate change: uncovering the adaptive significance. BioEssays. 2020 Nov;42(11):2000146.

www.ingramcontent.com/pod-product-compliance
Lightning Source LLC
Chambersburg PA
CBHW050010230526
45465CB00003BB/1351